W9-AUY-540

Woodlawn Library

IN MEMORY OF
George A. Thompson

WATER

by Robin Nelson

first step nonfiction

Lerner Publications Company · Minneapolis

We live on Earth.

Earth is made of
different things.

Earth is made of rocks, soil,
gases, and water.

Most of Earth is covered
by **water.**

Water can be a **liquid.**

Water can be a **solid**
called ice.

Water can freeze into ice.

Then ice can melt
into water.

People and animals need
water to live.

Plants need water to live.

We find water in **streams.**

Water is in lakes.

We find water in rivers.

Water is in **glaciers.**

We find water in oceans.

Water is found on Earth.

Water Experiment

This experiment may get messy, so ask a grown-up first.

Materials Needed:

- 2 T of water
- food coloring
- paper towel
- wax paper

Water moves up a plant.

Directions

Add a drop of food coloring to the water. Next, tear off a large piece of wax paper and lay it on the table. Slowly pour the water onto the middle of the wax paper. Water usually sticks to itself, so it should have made a small puddle. Hold the paper towel over the puddle with the edge touching the water. The way water moves in the paper towel is the same way water moves in a plant's stem.

Wild Water Facts

189 inches (480cm) of snow fell in one 1959 storm in Mount Shasta, California.

We have the same amount of water on Earth that we did millions of years ago. This means that the water from your faucet could contain some of the same water that dinosaurs drank.

Frozen water is lighter than liquid water. This is why ice floats in water.

A person may use 80–100 gallons of water a day. At home, flushing the toilet uses the most water.

At Mount Waialeale on Kauai, Hawaii, it rains 350 days a year.

Chemicals on the ground or in the air can pollute water. This can destroy animal homes.

All living things need water to survive.

Glossary

 glaciers – very large pieces of ice

 liquid – something that you can pour

 solid – something that has a definite shape

 streams – thin paths of water

 water – clear liquid with no taste

Index

The photographs in this book are reproduced through the courtesy of: © Jeff Greenberg, cover; © Jose Luis Pelaez, Inc./CORBIS, p. 2; © Allison Wright/CORBIS, p. 3; PhotoDisc Royalty Free by Getty Images, pp. 4, 12; NASA p. 5; © Norbert Schaefer/CORBIS, p. 6; © Trinity Muller/ Independent Picture Service, p. 7; © Michael T. Sedam/CORBIS, p. 8; © Greg Vande Leest/Photo Agora, p. 9; © Liba Taylor/CORBIS, p. 10; © Roy Morsch/CORBIS, p. 11; courtesy British Tourism Authority, p. 13; © Richard Hamilton Smith/CORBIS, p. 14; © Enzo & Paolo Ragazzini/CORBIS, p. 15; Courtesy of Cultural and Tourism Office of the Turkish Embassy, p. 16; © CORBIS, p. 17.

Lerner Publications Company
A division of Lerner Publishing Group
241 First Avenue North
Minneapolis, MN 55401 U.S.A.

Website address: www.lernerbooks.com

Library of Congress Cataloging-in-Publication Data

Nelson, Robin, 1971–
    Water / by Robin Nelson.
       p.   cm. — (First step nonfiction)
    Includes index.
    ISBN: 0–8225–2600–X (lib. bdg. : alk. paper)
    1. Water—Juvenile literature. I. Title. II. Series.
GB662.3.N456 2005
551.48—dc22   2004020784

Manufactured in the United States of America
1 2 3 4 5 6 – DP – 10 09 08 07 06 05